# GRASSHOPPERS

by Jane Dallinger

Photographs by Yuko Sato

A Lerner Natural Science Book

Lerner Publications Company ▪ Minneapolis

**Sylvia A. Johnson, Series Editor**

*Translation by Joe and Hiroko McDermott*

*The publisher wishes to thank Jerry W. Heaps,
Department of Entomology, University of Minnesota,
for his assistance in the preparation of this book.*

LIBRARY OF CONGRESS CATALOGING IN PUBLICATION DATA

**Dallinger, Jane.**
 Grasshoppers.

 (A Lerner natural science book)
 Adapted from Migratory locusts by H. Oda, originally
published under title: Tonosama batta.
 Includes index.
 SUMMARY: Describes the life cycle of grasshoppers
and discusses why some of these insects cause extensive
damage to crops.

 1. Locusts—Juvenile literature. [1. Grasshoppers.]
I. Satō, Yūkō, 1928- II. Oda, Hidetomo. Tonosama
batta. English. III. Title. IV. Series: Lerner natural
science book.

QL506.D34          595.7′26          80-27806
ISBN 0-8225-1455-9

This edition first published 1981 by Lerner Publications Company.
Revised text copyright © 1981 by Lerner Publications Company.
Photographs copyright © 1974 by Yuko Sato.
Adapted from MIGRATORY LOCUSTS. Text copyright © 1974 by
Hidetomo Oda and photographs copyright © 1974 by Yuko Sato.
English language rights arranged by Japan UNI Agency, Inc.,
for Akane Shobo Publishers, Tokyo.

International Standard Book Number: 0-8225-1455-9
Library of Congress Catalog Card Number: 80-27806

1  2  3  4  5  6  7  8  9  10  90  89  88  87  86  85  84  83  82  81

# A Note on Scientific Classification

The animals in this book are sometimes called by their scientific names as well as by their common English names. These scientific names are part of the system of **classification**, which is used by scientists all over the world. Classification is a method of showing how different animals (and plants) are related to each other. Animals that are alike are grouped together and given the same scientific name.

Those animals that are very much like one another belong to the same **species** (SPEE-sheez). This is the basic group in the system of classification. An animal's species name is made up of two words in Latin or Greek. For example, the species name of the lion is *Panthera leo*. This scientific name is the same in all parts of the world, even though an animal may have many different common names.

The next smallest group in scientific classification is the **genus** (GEE-nus). A genus is made up of more than one species. Animals that belong to the same genus are closely related but are not as much alike as the members of the same species. The lion belongs to the genus *Panthera*, along with its close relatives the leopard, *Panthera pardus*, the tiger, *Panthera tigris*, and the jaguar, *Panthera onca*. As you can see, the first part of the species name identifies the animal's genus.

Just as a genus is made up of several species, a **family** is made up of more than one genus. Animals that belong to the same family are generally similar but have some important differences. Lions, leopards, tigers, and jaguars all belong to the family Felidae, a group that also includes cheetahs and domestic cats.

Families of animals are parts of even larger groups in the system of classification. This system is a useful tool both for scientists and for people who want to learn about the world of nature.

Grasshoppers are insects famous for their big appetites. Grasshoppers eat plants, and sometimes a group of these insects will eat a farmer's entire crop.

Grasshoppers are also known for their jumping ability. Their long, powerful hind legs help them move quickly over the ground in big hops. The name *grasshopper* was probably given to them because they often can be seen in the grass, hopping around or enjoying a meal.

Some types of grasshoppers live peacefully with other animals and with plant life. They don't usually cause a lot of damage to crops. Other types of grasshoppers are very destructive. They travel in large groups that can destroy every plant throughout miles and miles of countryside. They have even been known to rip into shreds any laundry left hanging out to dry.

As you read this book, you will learn more about how grasshoppers live. You will also learn why some of them cause so much damage.

Grasshoppers live throughout the world. Some are found in rocky, mountain areas. A few species live in desert regions. But most prefer grasslands, where there is plenty of food, water, and sunshine.

Grasshoppers are usually quiet at night and active during the day. As with all insects, their bodies must be warm in order for them to be active. During the coolness of night they climb onto plants and rest quietly without moving. When the first rays of sunshine reach them in the morning, they begin to stir. After warming up, they start eating.

Grasshoppers will eat almost any kind of plant when necessary. But various types of grass and the green parts of flowering plants are their favorite foods.

Young grasshoppers, called **nymphs**\* (NIMFS), need a great deal of food to help them grow. In one day a nymph will eat food that weighs twice its own body weight. Imagine how much you would have to eat to equal this! If you weigh 60 pounds (27 kilograms), you would have to eat 120 pounds (54 kilograms) of food each day to eat as much as a nymph eats.

\*Words in **bold type** are defined in the glossary at the end of the book.

Some grasshoppers are found in rocky, mountainous regions (*below*), but most live in areas where there is plenty of grass (*right*).

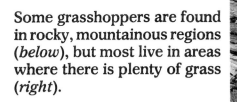

Grasshoppers belong to several different insect families. Some are members of the family Tetrigidae. These grasshoppers are also called pygmy **locusts**. Two pygmy locusts are shown on the next page in the picture on the top left. Pygmy locusts often live near ponds or streams. They are the smallest grasshoppers. As adults pygmies are about half an inch (1.25 centimeters) long.

Another family of grasshoppers is known as Acrididae. The picture on the top right shows a member of this family. This is a young insect whose wings have not developed yet. Some adults are 3 inches (7.5 centimeters) or more in length.

Grasshoppers in the family Acrididae are often called short-horned grasshoppers because they have short **antennae** (an-TEN-ee), or feelers. The white arrow in the picture (top right) points to the grasshopper's antennae. Most of the grasshoppers you will read about in this book are of the short-horned type.

An adult grasshopper of the family Tettigoniidae is shown in the picture on the bottom left. The white arrow points to its antennae. This type of insect is often called a long-horned grasshopper because its antennae are so long.

Several other insects are closely related to grasshoppers. Katydids belong to the same family (Tettigoniidae) as long-horned grasshoppers. Crickets (bottom right picture) are also relatives of the long-horns. Mantises (shown on page 37) and cockroaches are two other close relatives of the various types of grasshoppers.

## THE BODY OF A GRASSHOPPER

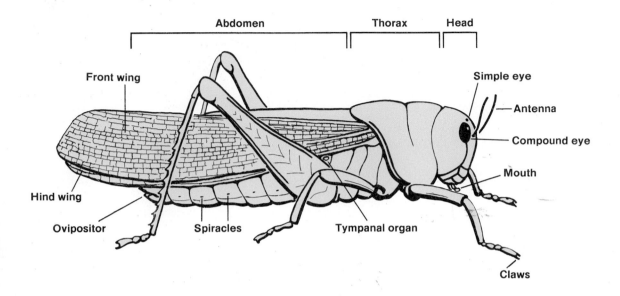

Like all insects, grasshoppers have three main body sections: the head, the thorax, and the abdomen.

On a grasshopper's head are two antennae and five eyes. The antennae are covered with fine hairs that are very sensitive to touch. The two large eyes, called compound eyes, are used to see movement and objects at a distance. The three smaller eyes, called simple eyes, respond to changes in light. On the back of the head is an area called the **ampulla** (am-PUHL-uh). Later in this book you will learn how grasshoppers use the ampulla when they hatch and when they shed their skin.

The thorax lies just behind the head. A grasshopper's two pair of wings are attached to the thorax. The large, delicate hind wings fold under the stronger front wings when a grasshopper isn't flying. As adults, most grasshoppers have wings. But grasshoppers of some short-horned species have very weak wings, and others have no wings at all.

All grasshoppers have three pairs of legs attached to the thorax. Many of the insects also have a **tympanal** (TIM-puh-nahl) **organ** on each side of the thorax. Tympanal organs serve as ears for grasshoppers.

The abdomen of a grasshopper lies behind the thorax. Females have an **ovipositor** (oh-vih-PAHZ-ih-tur) at the end of the abdomen. The ovipositor is used to lay eggs and to dig holes in the ground so that the eggs can be buried. All grasshoppers have breathing holes, called **spiracles** (SPY-reh-kuhls), on their abdomens.

A grasshopper has 10 spiracles along each side of its abdomen. Air is brought into the body through the 4 spiracles nearest the head. Air leaves the body through the 6 spiracles farthest from the head. If you watch a grasshopper closely, you can see its abdomen move out and in very regularly as the insect breathes.

Sometimes grasshoppers accidentally land in water while hopping or flying. Most have to get out of the water quickly. Otherwise, water would enter their bodies through the spiracles and they would drown. Only pygmy locusts and some long-horned grasshoppers can close their spiracles so water cannot get in.

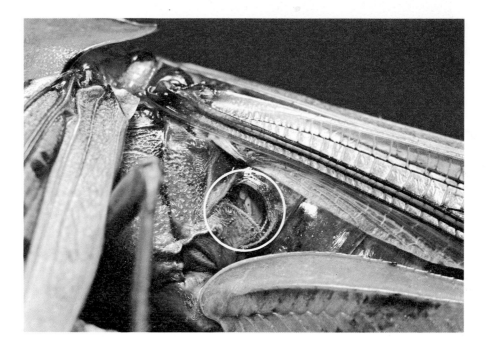

Many species of grasshoppers have tympanal organs that serve as their ears. In short-horned grasshoppers, a tympanal organ is found on each side of the thorax, just above the hind leg. A tympanal organ is marked by a white circle in the picture above. In long-horned species the tympanal organs are found on the front legs.

Some species of short-horned grasshoppers do not have tympanal organs. Hearing probably does not play an important role in their lives. Some long-horned species have very sensitive tympanal organs. They can hear a wider variety of sounds than humans can. Hearing is probably very important to these grasshoppers.

The sounds that grasshoppers themselves make play an important part in their lives. Male grasshoppers make sounds by using a method called **stridulation** (strij-eh-LAY-shun), or rubbing two parts of the body together. The short-horned male on the opposite page is stridulating by moving a hind leg over a front wing. Males of some short-horned species move their front wings against their hind wings to make sounds. Special structures on the areas being rubbed together help create the grasshoppers' songs.

A long-horned grasshopper makes sounds by rubbing the edge of one front wing along a special ridge found on the bottom of the other front wing. This is the same method of stridulation used by crickets. The black cricket pictured above is stridulating, and so is the long-horned grasshopper shown on page 7 (bottom left).

Female grasshoppers and males of some short-horned species do not stridulate.

Stridulation serves three main purposes. It helps grasshoppers of the same species to find each other and stay together. Each species has its own songs, which are recognized by the members of the species. Stridulation also protects a male's territory by warning other males to stay away. Finally, males attract females with their songs.

You are most likely to hear grasshoppers stridulating in late summer and early fall. This is their mating season. During this season, the males sing as they wait for females to approach.

When a male grasshopper is approached by a female, he strokes her with his antennae. Then he climbs on top of her, and they mate. The picture on the opposite page shows two grasshoppers mating. Notice that the male is smaller than the female. This is usually true for grasshoppers.

After mating, the two grasshoppers go their separate ways. The female carries on her normal activities for about a week. Then she is ready to lay her eggs.

Before a female grasshopper lays her eggs, she digs a burrow for them. She uses the ovipositor on the end of her abdomen as a drill. The grasshopper pushes her ovipositor and her abdomen into the ground (opposite). Then she packs the soil to each side with the ovipositor. Slowly, she drills farther and farther into the ground. Eventually her abdomen may stretch down until it is two or three times its normal length.

When the burrow has been completed, the female lays her eggs. She surrounds each egg with a thick layer of foam. This foam also lines the entire burrow. As the foam hardens, it forms a protective covering for the eggs.

A grasshopper may lay anywhere from 20 to 120 eggs in one burrow. All of these eggs and the foam that covers them are called an **egg pod**. It often takes about an hour for the egg pod to be completed. Once a female finishes making the pod, she pushes dirt over the opening of the burrow (above). This helps protect the eggs from enemies that might eat them. The picture on the opposite page shows an egg pod buried under about one inch (about 2.5 centimeters) of dirt. The egg pod has been cut in half so you can see inside of it. Each of the yellow objects that look like jelly beans is one egg.

Most female grasshoppers will lay more eggs after making their first egg pod. An average female makes about 8 egg pods during the 4 to 12 months that she lives. Some grasshoppers have been known to make 25 egg pods, but this large number is very unusual.

Most grasshopper eggs are laid in the fall. The eggs develop for a short time, but when the temperatures drop as winter comes, they stop growing. When temperatures rise with the beginning of spring, development starts once more. Many of the eggs are ready to hatch just after the first spring rains.

Inside its egg, a developing grasshopper is covered by a protective skin. In order to begin hatching, the grasshopper first bites through this skin. Then it spits out a special liquid that softens the egg shell around its head. Finally, it begins to push its way out.

A grasshopper can't use its legs during hatching because they are wrapped tightly inside the protective skin. The insect begins to hatch by using the ampulla on the back of its head. This area fills with blood and swells up. It pushes against the egg shell, causing the shell to rip open. Then the grasshopper moves out of the egg by pushing and wriggling like a worm.

A newly hatched grasshopper still has more work to do after it leaves the egg. It must move through the outer layer of the egg pod and a layer of dirt in order to reach the surface of the ground.

Once a grasshopper wriggles to the top of the ground (above), it must shed the protective skin that surrounds its legs and antennae. This is called **molting**. The grasshopper molts in much the same way that it hatched from its egg. The ampulla swells up with blood and pushes against the skin. The skin rips, and the grasshopper crawls out of it. In this stage of its life, a young grasshopper is called a nymph. Nymphs look very much like adult grasshoppers, except that they have no wings.

The nymphs on the following pages have recently shed their skins. Notice that their antennae still bend forward over their heads, and their legs droop down near their bodies.

A nymph's body is moist and soft after molting. It takes a while for the legs and antennae to dry and harden so that they can be used. The legs of the nymph in the picture above are springing into place. But its antennae are still bent forward over its head.

A short time later, the nymph is ready to go searching for its first meal (opposite). Its antennae and legs are in their proper positions. The nymph can now use its legs to move along the ground. As it moves away, it leaves behind the old skin it just molted. The skin has shrunk into a white ball. In places where many grasshoppers live, the ground is covered with these white balls every spring.

Nymphs move along the ground by crawling or by hopping. Since they don't yet have wings, they are unable to fly. If a nymph is frightened, it makes a long leap that may be 50 to 100 times longer than the length of its body (opposite).

Nymphs spend much of the daytime eating (left). Their shapes and their green or brown coloring make them look like the plants they eat. This protects them from being seen and caught by birds or other enemies. You may find it hard to see the long, slender nymph in the picture on the right. It looks much like the blade of grass it is eating.

As a nymph grows, it becomes too large for its skin. The skin of a grasshopper isn't soft, like human skin. It is a hard, protective shell, and it can't stretch. Because of this, a growing nymph must molt about once a week.

When nymphs are ready to molt, most of them climb onto grass or other plants. After climbing up a plant, a nymph turns so that it hangs upside down (opposite). Moving forward slowly, it slides out of the old skin. Pygmy locusts molt in a slightly different way. They stay on the ground instead of hanging from a plant.

Just after molting, a nymph usually remains in one place for a while. The warm sun shines down on the new skin, drying it. In a short time the young grasshopper is able to move about and begin eating.

Grasshoppers of most species shed their skins a total of six times. This includes the first molt, just after hatching. A nymph is about half its adult size after the fourth molt. At this time, small stubs appear on its back. These are the beginnings of wings. The wings are half grown by the fifth molt and are almost ready to use by the sixth molt. A nymph becomes an adult grasshopper after the sixth molt. An adult doesn't grow very much, so it keeps the same skin for the rest of its life.

The grasshopper shown on the next four pages is going through its sixth molt.

The sixth and last molt takes place much like earlier ones. Holding onto a blade of grass with its legs, the grasshopper begins to shed its skin. The ampulla swells up with blood and pushes against the old skin until the skin tears. Gently, the grasshopper pulls its head and antennae out of the skin through the hole (top).

Next, the grasshopper drops down slowly. The rest of its body slides out of the old skin (middle). Finally, it hangs from the skin, connected to it only at the end of the abdomen (bottom).

It takes about an hour for the grasshopper to reach this stage. The insect then rests quietly for a few minutes. Getting free from the skin took a great deal of effort, so the grasshopper is tired.

After resting, the grasshopper bends forward. It uses its front feet to grab onto the old skin it has just shed (top). Then the grasshopper pulls itself up until it can reach the blade of grass. Holding onto the grass with its front legs (bottom), the insect breaks free from the old skin.

If you go back to the first picture in this series, you will see that before molting, the wings look like short stubs. Just before the last molt, the wings are fully developed. But they are folded tightly inside the skin, so they look very short. After this molt, the wings unfold and stretch to their full length.

During molting, grasshoppers swallow a lot of air. The air and extra blood are now pumped to the surface of the body. This helps the wings to unfold.

**This grasshopper is resting as its new skin hardens and its wings stretch out to dry.**

The wings and skin of a grasshopper take two hours or longer to dry and harden after molting. Once this drying is finished, the grasshopper is ready to begin its adult life as a flying insect.

34

It is late spring by the time grasshoppers become adults. Throughout the summer their bodies continue to develop as they prepare for mating season in the fall. During this time, one of the biggest dangers for a grasshopper is the threat of being eaten by an enemy.

Grasshoppers have many enemies. The garden spider above has just caught a grasshopper in its web. The spider will eat the insect after paralyzing it with a poisonous bite. On the opposite page, a praying mantis holds a grasshopper by using its front legs, which are specially designed to capture insects.

Grasshoppers are also eaten by many types of birds. The grasshopper on this page was caught by a shrike and left hanging on some dried grass. Perhaps the shrike was frightened away by one of its own enemies before it could eat the grasshopper.

Other animals that hunt grasshoppers include monkeys, lizards, snakes, and mice. One type of mouse is even called a grasshopper mouse because grasshoppers are its main source of food. Some flies eat grasshoppers, too. Baby flies burrow under the skin of a grasshopper and live there, using the insect's body as food. Sometimes the grasshopper continues to live a normal life, but in some cases it dies.

**This green lizard is one of the many animals that hunt and eat grasshoppers.**

Many enemies eat grasshopper eggs. Young beetles will crawl into an egg pod and live there, using the eggs for food. Other adult insects lay their own eggs inside a grasshopper's egg pod. As the babies of these insects grow, they eat the grasshopper eggs.

It is important to remember that animals who kill grass-hoppers are usually being helpful. Grasshoppers eat plants, and if too many of these insects gather in one area they can become very destructive.

Now and then a very large group of grasshoppers develops in one area. Why this happens is not always known. Hot, dry weather can sometimes be the cause. In such weather, female grasshoppers lay many eggs. At the same time there may be fewer natural enemies around to eat the eggs or the nymphs that hatch from them. Whatever the cause, when a large number of nymphs live together in a small area (opposite), some unusual events occur.

Nymphs that live in crowded conditions are called **gregarious** (greh-GAIR-ee-us) **nymphs.** They develop differently than grasshoppers do under normal conditions. The crowding affects certain chemicals in their bodies. These chemicals, known as hormones, change and cause the gregarious nymphs to become darker than usual in color. Compare the gregarious nymph in the picture on the right with the normal nymphs, also called **solitary nymphs**, on the left.

Changes in their hormones cause gregarious nymphs to develop quickly, too. They grow very fast and go through their molts sooner than solitary nymphs.

Gregarious nymphs act differently than solitary nymphs. Gregarious nymphs are restless. They move around a great deal. When gregarious nymphs move, they often copy the actions of their neighbors. If one gregarious nymph jumps into the air, the others nearby will jump too. Solitary grasshoppers don't copy each other in this way.

Gregarious nymphs tend to form large bands that migrate, or travel long distances. In fact, **migratory locust** is another name for this type of grasshopper. The nymphs are most likely to begin migrating at midday.The warm temperatures make them very active. A few of the nymphs will start to jump about. Then their neighbors copy them. Soon the entire band is hopping together in one direction. Once the band begins to migrate, it usually keeps moving until the nymphs are quieted down by the cooler temperatures of evening.

During the day, bands of migrating nymphs sometimes meet. When this happens, they join and become one large group that marches along the ground. In the United States during the 1800s, one migrating group was reported to be 23 miles (37 kilometers) wide and 70 miles (112 kilometers) long. Since the insects in such a group must eat each morning, they usually destroy much plant life during such a migration.

Migratory locusts stay together even after they develop wings and are able to fly. Just after the sixth molt, migratory locusts take short flights together each day. After several days of practice on these shorter flights, they go on a journey that may last as long as three days. They seem to come down only when cold temperatures force them down, or when they become very hungry. Upon landing, they destroy everything in sight — grass, farmers' crops, the bark of trees, even clothes, curtains, and other things made of cloth.

The picture on the left shows a swarm of locusts flying through a village in Africa. Africa is one of many areas in the world that have had problems with migratory locusts.

45

Migratory locusts caused serious crop damage in the United States many times during the 1800s. In the first part of the 1900s, chemicals were used to kill the insects. But these chemicals proved to be harmful to humans, and many farmers stopped using them. As a result, the grasshopper population rose again. In the 1970s, farmers in the Midwest began to have problems with grasshoppers eating their crops.

Scientists are looking for safer, more natural methods to keep down the grasshopper population. One thing that seems to help is plowing the fields in the late fall and early spring. This exposes the grasshopper eggs so that they can be easily found by animals that eat them. Scientists hope that such natural methods of control will some day make it possible for grasshoppers to live in harmony with other forms of life.

# GLOSSARY

**ampulla**—an area on a young grasshopper's head used to break through the egg shell when the grasshopper hatches. The ampulla is also used to break through the old skin during molting.

**antennae**—sense organs on the heads of grasshoppers that are very sensitive to touch.

**egg pod**—a collection of grasshopper eggs covered by foam

**gregarious nymphs**—young grasshoppers that develop under crowded conditions and often travel together in groups

**locust**—a name sometimes used instead of *grasshopper*, especially for insects that cause damage to crops

**migratory locusts**—grasshoppers that travel long distances in large groups

**molting**—shedding an old skin to make way for a new one

**nymphs**—young grasshoppers before the sixth molt has taken place and the wings have completely developed

**ovipositor**—a structure at the end of a female grasshopper's abdomen used to lay eggs and to dig a burrow for them

**solitary nymphs**—young grasshoppers that develop under normal conditions and do not migrate

**spiracles**—breathing holes on a grasshopper's abdomen

**stridulation**—a method of making noise by rubbing two body parts together

**tympanal organ**—a structure on a grasshopper's thorax or legs that receives sound

# INDEX

Acrididae (family), 8
ampulla, 10, 22, 23, 32
antennae, 8, 10, 23

body of grasshopper, 10

cockroaches, 8
compound eye, 10
controlling grasshoppers, 46
crickets, 8, 15

damage caused by grass-
    hoppers, 5, 43, 45-46
development of grass-
    hoppers, 31-35

egg pod, 20, 23
eggs, 19, 20, 22, 39, 40, 46
enemies, 29, 36-39
environment of grass-
    hoppers, 6

families of grasshoppers, 8
food, 6, 29

gregarious nymphs, 40, 43

katydids, 8

locusts, 8
long-horned grasshoppers,
    8, 12, 13, 15

mantises, 8, 36
mating, 16, 26, 36
migration, 43-46
migratory locusts, 43-45
molting, 23, 31, 33

nymphs, 6, 23, 26-34, 40,
    43

ovipositor, 11, 19

pygmy locusts, 8, 12, 31

relatives of grasshoppers, 8

short-horned grasshoppers,
    8, 13, 15
solitary nymphs, 40, 43
spiracles, 11, 12
stridulation, 15, 16

Tetrigidae (family), 8
Tettigoniidae (family), 8
tympanal organ, 11, 13